I am STEM

a STEM Fun Book

By: Corell Oglesby

ISBN: 979-8-9859187-2-4

To my family:

Thank you for believing in me.
You inspired me to do greater
and I am grateful for the push.

I am an Inventor
Alice H Parker

Born in 1895, Alice H. Parker was as an American inventor and a cook. She grew up in Morristown, New Jersey. She invented the first central heating system used in homes.

Other Important Inventors
Marie Van Brittan Brown

"We must believe that we are gifted for something and that this must be attained"

Marie Van Brittan Brown was an innovator and a nurse. She was born in Queens, New York in 1922. Marie, alongside her husband, invented the video home security system".

Mary Beatrice Davidson Kenner

Mary Beatrice Davidson Kenner has the record for the most patents and inventions awarded to a Black woman in the U.S government. She began inventing at the age of six.

Mary was born in 1912 in Charlotte, North Carolina and later moved with her family to Washington, D.C. Mary's most famous inventions were the sanitary belt and the toilet paper holder.

Lyda D. Newman

Born in 1865, Lyda D. Newman was the third black woman to receive a patent. She was born in Ohio but spent the majority of her life in Manhattan, New York.

Lyda invented the first hairbrush with synthetic bristles.

Sarah Elisabeth Goode

Born in 1855, Sarah Elisabeth Goode invented the first folding cabinet bed. She grew up in Ohio but moved to Chicago, Illinois at the end of the Civil War.

A high school opened in 2012 in her honor called Sarah E. Goods STEM Academy.

Ellen Eglin

Ellen Eglin was a housekeeper and a government clerk. She was born in 1849 and grew up in Washington DC. She invented the clothes-wringer in 1888.

Judy Woodward Reed

Born in 1826 in Charlottesville, Virginia, Judy Woodford Reed was the first African American female inventor to receive a patent. Judy invented the dough kneader and roller.

Bessie Blount Griffin

"A black woman can invent something for the benefit of humankind."

Bessie Blount Griffin was a forensic handwriting analyst, nurse, and a physical therapist who worked with injured soldiers back in World War 2. She was born in rural Virginia in 1914 and later settled in New Jersey. Bessie invented the electric feeding tube and taught amputee veterans to write with their teeth and their feet.

Valerie Thomas

"Hobbies are for wimps who don't follow their passion."

Born in 1943 in Maryland, Valerie Thomas is an American scientist and invented the illusion transmitter, for which she received a patent in 1980. Valerie was responsible for developing the digital media formats image processing systems used in the early years of the Landsat program at NASA and later used in motion pictures and 3D movies.

Marian Croak

"You don't have to be a victim of trouble. You can rise above problems and fix them."

Born in 1955 in New York City, Marian Croak is a prolific inventor and engineer in areas of information technologies. She has over 200 patents to her name. Marian is best known for developing Voice Over Internet Protocols (VoIP), a technology that converts your voice into a digital signal.

Gladys Mae West

Born in 1930, Gladys Mae West is an American mathematician known for her contributions to the mathematical model of the Earth. She was born in Sutherland, Virginia. Gladys is acclaimed for having contributed to the creation of the Global Positioning System (GPS).

I am a Neurosurgeon
Alexa Canday

"If you do good work, the rest doesn't matter."

Alexa Canday is a retired American medical doctor specializing in pediatric neurosurgery. She was born in Lansing, Michigan in 1950. She was the first black woman to become a neurosurgeon.

I am an Astronaut
Mae Carol Jamison

"Never be limited by other people's limited imaginations."

Born in 1956, Mae Carol Jamison was the first black woman to travel to space aboard the Endeavor. She was born in Alabama and raised in Chicago. Mae studied engineering, physics, and a retired NASA Astronaut.

I am a Scientist
Katherine G Johnson

"Like what you do, and then you will do your best."

Katherine G. Johnson was a mathematician and the first woman to ever work as a NASA scientist. She was born in White Sulphur Springs, West Virginia in 1918. She has a computational research facility named in her honor called the "Katherine G. Johnson Computational Research Facility."

I am a Mathematician
Dorothy Johnson Vaughn

"I changed what I could, and what I couldn't, I endured."

Dorothy Johnson Vaughn was an American mathematician and a human computer who worked at NACA and later at NASA Langley Research Center. She was born in Kansas City, Missouri in 1910 and later moved to Morgantown, West Virginia. Dorothy prepared for the introduction of machine computers by teaching herself and her staff the Fortrana programming language.

Mary Winston Johnson

"Every time we have a chance to get ahead they move the finish line. Every time."

Born in 1921, Mary Winston Jackson was an American Mathematician and aerospace engineer. She was born and raised in Hampton, Virginia.

Mary was the first African-American female engineer to work at NASA.

I am a Surgeon
Dorothy L. Brown

"I am proud to be a role model, not because I have done so much, but to say to young people that it can be done."

Dorothy Lavinia Brown was a surgeon, a legislator, and teacher. Dorothy was born in Philadelphia, Pennsylvania in 1914. She became the first African American female appointed to a general surgery residency in 1971.

I am a Doctor
Rebecca Lee Crumpler

"I early conceived a liking for, and sought every opportunity to relieve the sufferings of others."

Rebecca Lee Crumpler was born in 1831. She was a physician, nurse, and author. She was born in Christiana, Delaware, raised by her aunt in Pennsylvania and later moved to Charlestown, Massachusetts. Rebecca became the first African American woman to become a doctor of medicine in the US.

I am a Pyschophysiologist
Patricia S. Cowings

"Scientists are eternal students. We ask questions for a living."

Patricia S. Cowings is an aerospace psychophysiologist and was an alternate for a space flight in 1979. She was born in and raised in The Bronx, New York City in 1948.

Patricia developed and patented a physiological training system called Autogenic Feedback Training Exercise (AFTE), which enables people to learn voluntary self-control of up to 24 bodily responses in six hours.

I am an Opthamologist
Patricia Era Bath

"Do not allow your mind to be imprisoned by majority thinking. Remember that the limits of science are not the limits of imagination."

Patricia Era Bath was an ophthalmologist, laser scientist, innovative research scientist, and advocate for blindness prevention, treatment, and cure. She was born in Harlem, New York in 1942. Patricia was the inventor of laser cataract surgery called "Laserphaco Probe".

I am a Biochemist
Mary Maynard Daly

"Courage is like – it's a habitus, a habit, a virtue: you get it by courageous acts."

Born in 1921 in Queens, New York, Marie Maynard Daly, was the first woman in the United States to earn a Ph.D in chemistry in 1947. She was an American biochemist, researcher, and instructor.

I am a Physicist
Shirley Ann Jackson

"Do not be limited by what others expect of you, but confidently reach for the stars."

Born in 1946, Shirley Ann Jackson is an American physicist and the first African American woman to have earned a doctorate at the Massachusetts Institute of Technology (MIT) in 1973. She was born in Washington, D.C. Shirley also served as Chairman of the US Nuclear Regulatory Commission (NRC), becoming the first woman and first African American to hold that position in 1995.

I am a Nuclear Chemist
Claire Phelps

"Through my story, I want to encourage young people to build their own path and make discovery about the potential that lies within them."

Claire Phelps was born in Tennessee and graduated Tennessee State University in 2003. Soon after, she joined the US Navy and enrolled in the Navy's Nuclear Power School.

Claire Phelps was on the discovery team that found a new element, named Tennessine, the 117th element. She is the first African American woman involved with the discovery of a chemical element.

23

I am STEM
Vocabulary Building

Vocabulary

1. An **<u>inventor</u>** is a person that creates a process or thing.

2. A **<u>patent</u>** is a license to the exclusive rights to exclude others from creating, using, or selling an invention.

3. A **<u>synthetic</u>** item is created to mimic a natural, original item

4. **<u>Forensic</u>** is the use of science to investigate crimes.

5. **<u>Three Dimensional</u>** is the appearance that the depth, length, and height is believable.

6. **<u>VoIP</u>** stands for Voice over Internet Protocol which allows you to use the internet to make phone calls.

7. **<u>GPS</u>** is a satellite base application that assists with navigation.

8. A **<u>neurosurgeon</u>** is someone that specializes in the surgery of the nervous system, brain, and spinal cord.

9. An **<u>astronaut</u>** is someone that is trained or has traveled to space.

Vocabulary

10. A **scientist** studies and works in the world of science.

11. A **mathematician** studies and works in the world of math.

12. **Doctors** are healthcare professionals that medically study and treat humans, animals, plants, etc.

13. A **psychophysiologist** studies the physiological process of psychology.

14. An **ophthalmologist** is a doctor that treats your eyes.

15. **Biochemists** study chemical reactions in living organisms.

16. A **physicist** studies physics.

17. **Physics** is the study of matter.

18. **Matter** is a physical substance.

19. An **element** is pure substance and cannot be reduced to a simpler form.

20. An **electronic device** uses microchips, transistors or other components.

I am STEM
Coloring & Activity Pages

Match-up Puzzle

Match the vocabulary to the definition.

_____ Invention

_____ Patent

_____ Science

_____ Experiment

_____ Computer

_____ GPS

_____ Periodic Table

_____ Three Dimensional

_____ Data

_____ Astronaut

_____ Mathematician

_____ Mechanic

_____ Spaceship

_____ Planets

_____ Computer Science

a. the study of computers

b. a table of chemical elements

c. the study of the world and nature through observation and experimentation

d. a person that studies mathematics

d. someone who has traveled to space

e. a celestial body

f. a satellite-based system used for navigation

g. exclusive rights to an invention

h. the study of math and motion

i. the creation of something

j. an aircraft used to travel to space

k. information saved on a device

l. exploring and testing especially in a lab to make a determination

m. a device used to store and process data

n. technology that makes the depth, length, and height is believable

28

I can be a Scientist.

I can be a Technologist.

I can be an Engineer.

I can be a Mathematician.

Number Block Puzzle

Fill in the missing numbers

The missing numbers are integers between 0 and 5.

The numbers in each row add up to totals to the right.

The numbers in each column add up to the totals along the bottom.

The diagonal lines also add up the totals to the right.

1

					7
					15
					15
0					9
			2		6
	2		2		16
12	12	13	12	12	14

2

					14
	2				4
					12
					7
4			3		14
4					16
13	7	9	15	9	10

Number Block Puzzle Con't

3

					11
					15
1					13
1					8
	5			5	17
					16
10	16	6	20	17	13

4

					12
				5	19
					9
5					11
5					13
				1	10
14	7	15	13	13	6

Hidden Message Puzzle

This puzzle is a word search puzzle that has a hidden message in it.

First, find all the words in the list.
Words can go in any direction and share letters as well as cross over each other.
Once you find all the words, copy the unused letters starting in the top left corner into the blanks to reveal the hidden message.

You can do it.

```
M  Y  O  U  C  R  E  A  T  E  E  C  A  N  B
E  A  A  N  O  Y  T  H  I  C  N  G  J  V  S
F  N  T  T  G  T  U  A  N  O  R  T  S  A  C
O  K  C  H  F  C  A  E  Q  E  H  I  A  A  H
X  O  E  I  Q  G  I  G  W  M  U  I  X  Y  O
D  R  E  P  O  C  S  O  R  C  I  M  T  M  O
K  S  O  Z  S  N  Q  T  C  C  X  E  T  Y  L
J  S  P  T  Z  D  P  T  K  C  C  S  L  L  Z
D  B  N  Z  N  W  D  G  A  H  B  W  R  E  Y
L  D  F  E  A  E  V  Q  N  D  C  E  G  A  W
I  J  N  M  M  I  V  O  Z  R  Q  Z  O  R  E
U  D  X  E  O  O  L  N  N  P  O  M  P  N  O
B  R  A  T  Q  O  W  S  I  A  Y  W  R  N  C
I  U  I  S  G  R  E  E  N  I  G  N  E  U  B
Y  X  Y  Y  N  C  C  O  M  P  U  T  E  R  Y
```

astronaut	build	computer
create	doctor	engineer
inventor	learn	math
microscope	school	science
STEM	technology	women

___ ___ ___ ___   ___ ___ ___   ___ ___   ___ ___ ___ ___ ___ ___ ___ ___

Find the Word

```
K L R E L E M E N T S S J S R
S A T G B W C N U N H N L O F
Z W V L A N D I N O S A U R S
A S V T L I S S O F C W C T P
O Y E U S T N E M I R E P X E
P R E E O T P C M T S I J O S
U F U P A L D E G H P B E T N
I S P S O G H L B G W M S N W
B L L T E C A L Z I A E A F Y
C A X N I S S S C E T T C S D
W M L E L V U O W W O I O L S
R I B D N O L A R W F U A M N
L N L U E O W V C C M W Y F J
P A T T X Z F M C L I M A T E
N S T S I T N E I C S M Y U R
```

ANIMALS	ATOM	CAUSE
CELLS	CHEMICALS	CLIMATE
DINOSAURS	ELEMENTS	EXPERIMENTS
FOSSIL	GAS	LAB
LAWS	MASS	MICROSCOPE
SCIENTISTS	STUDENTS	TESTS
WATER	WEIGHT	

Find the Word

```
C T S F X A R M G L U T M S B
C H D R J Z N O K O G I F Y S
S B I K O G G U I C I X T Y V
V C R P N T S S X A I E O F F
Q S A I S N I E V C S L A U A
J M G N O C L N S S Q N C Q T
R O G T O B D R O W S S A P A
L O T T J Z E W K M R P E N D
T U W J C T N E E R C S T W S
B N T K U C O D I N G F I C K
W N N P E P K W U Z H I S Y U
A R M U R Y V O C T R L B F H
R O V I R U S N Z E F E E Y D
C O N X C C Z L E M M S W G P
S T C R O D M E M O R Y O J Q
```

buttons	bytes	chips
click	coding	computers
data	exit	files
keys	login	memory
monitors	mouse	password
print	scan	screen
virus	website	

Find the Differences

Find 5 Differences:

Create your own Puzzle

Draw a picture. Make sure you color it in. Cut out your puzzle pieces. Challenge someone to put your puzzle together.

Color with the number

2 + 2 =   Red   1 + 1 =   Green

1 + 2 =   Yellow   2 - 1 =   Blue

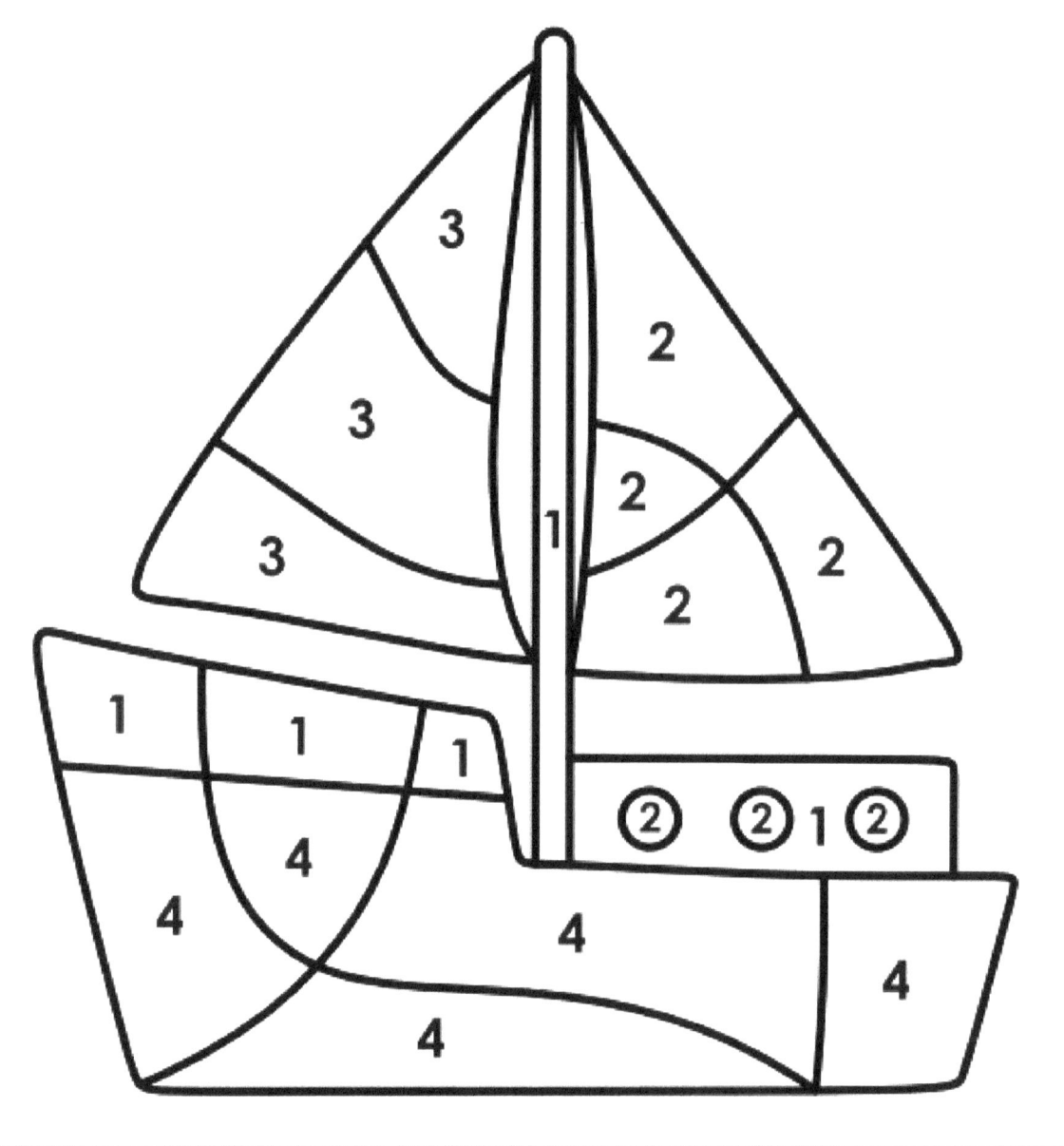

Let's Play in the Stars!

Planets Crossword

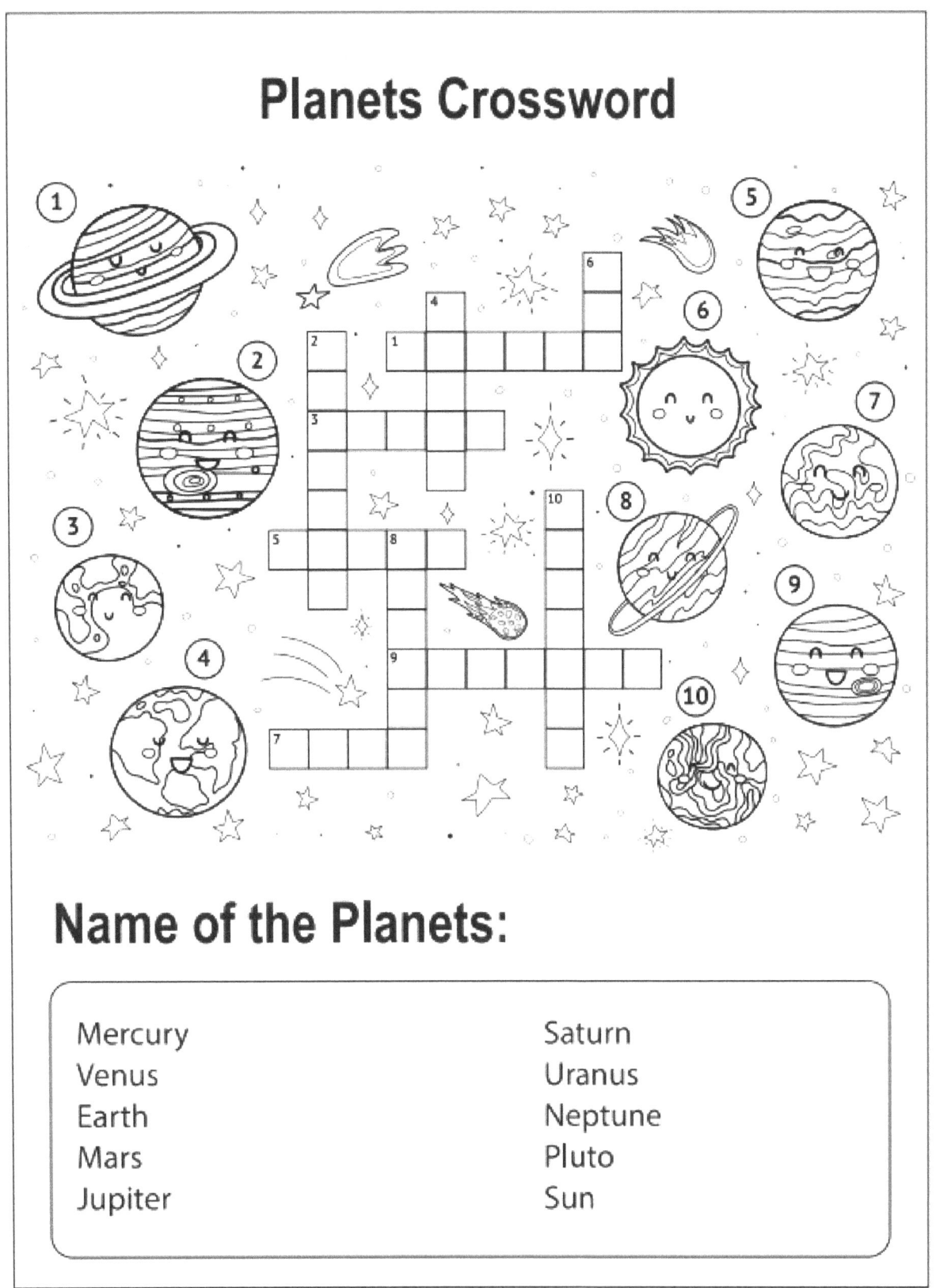

Name of the Planets:

Mercury   Saturn
Venus   Uranus
Earth   Neptune
Mars   Pluto
Jupiter   Sun

Connect the Dots!

Connect the dots and name your robot!

Name: _____

Connect the Dots!

Connect the dots and discuss.

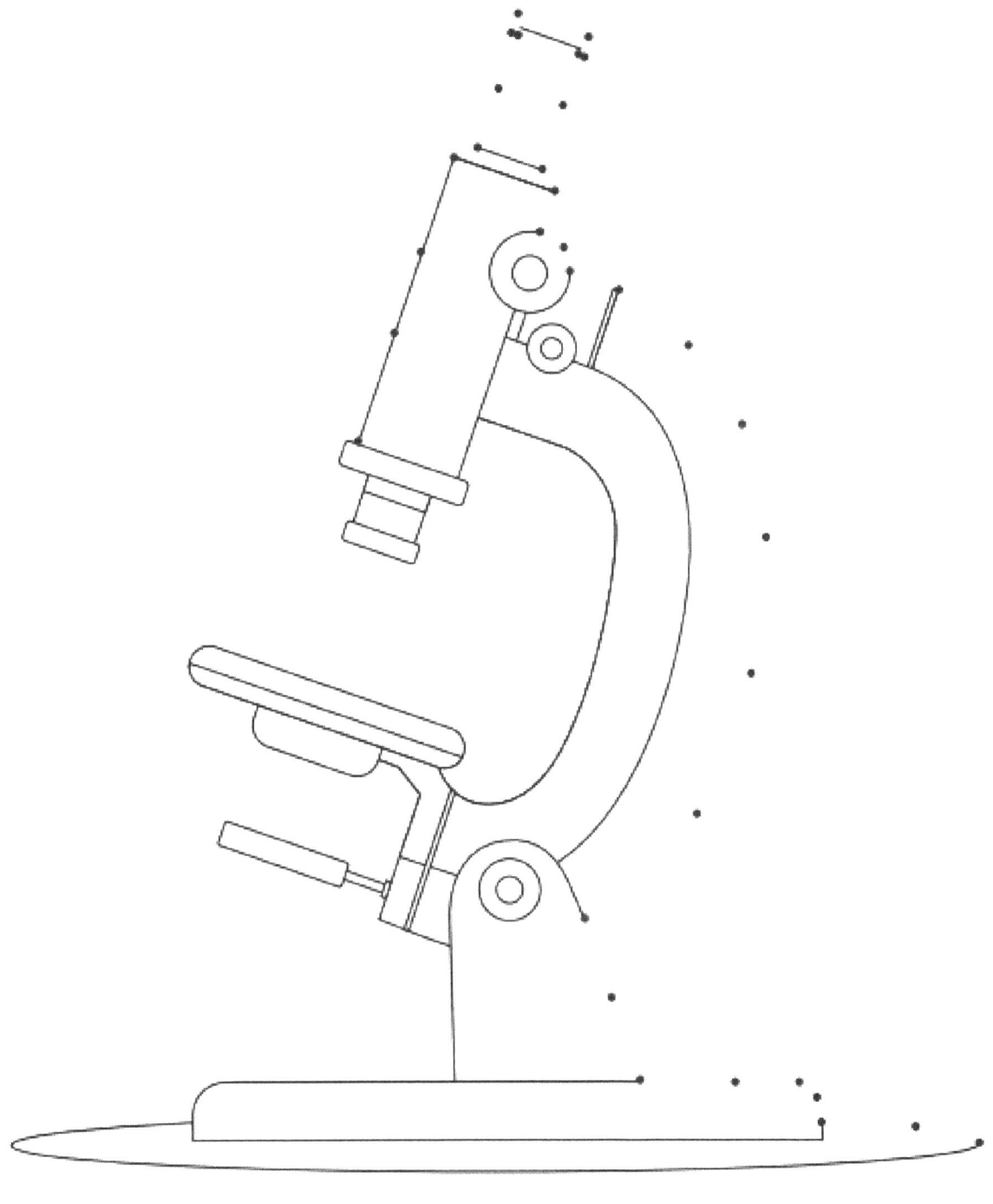

What is the item?
What can it do?

Find the lab

Help the scientist get to her lab.

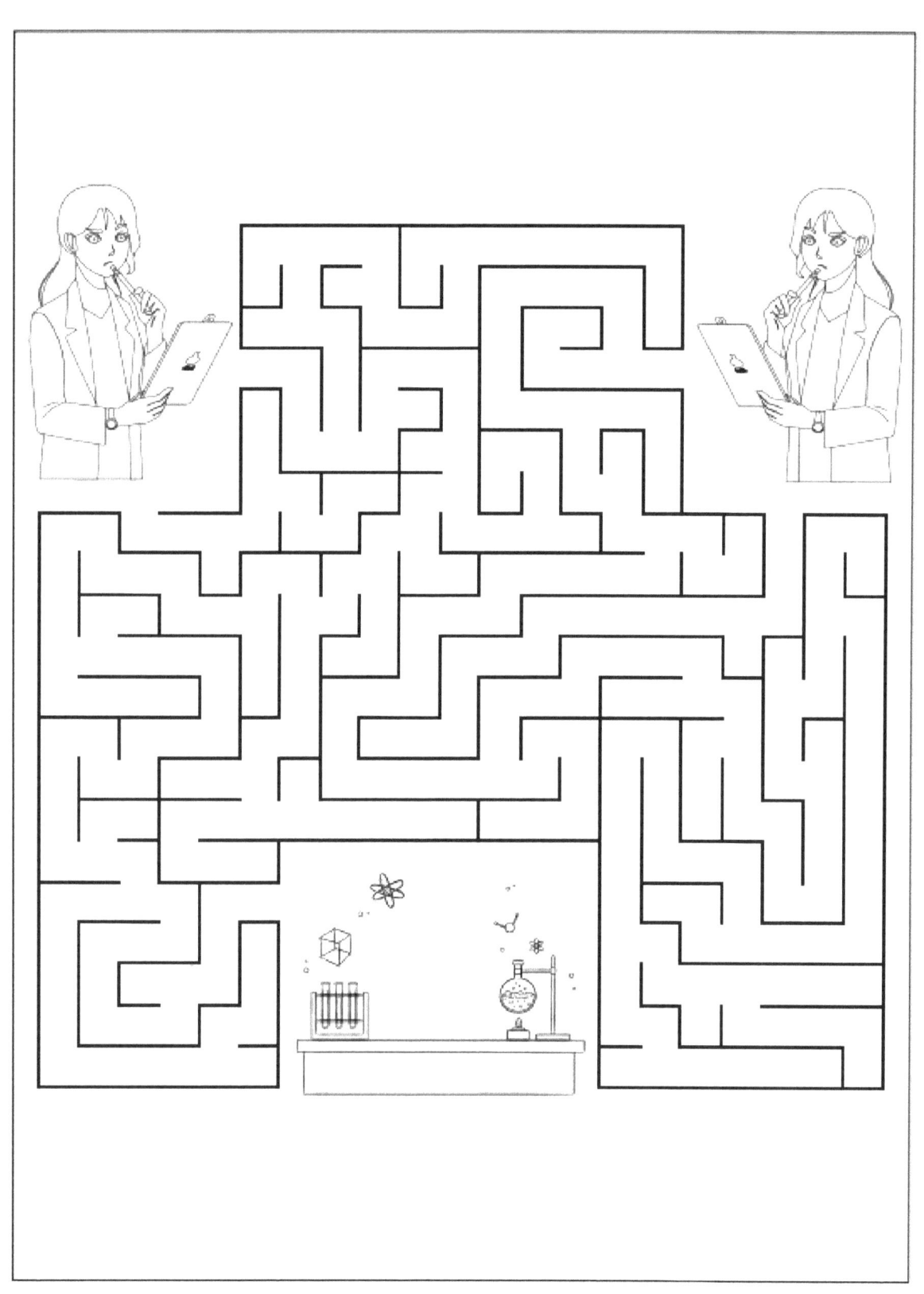

Let's Experiment!

STEM can teach your student how to be a team player, how to think logically, improve problem-solving ability, and enhance creative thinking. Here are a few STEM activities your elementary age student can explore. Make sure you engage in plenty discussion with your student. Have fun exploring!

Rainbow Walking Water

Tools needed:

a. 6 mason jars

b. 6 liquid food colors (red, yellow, and blue)

c. paper towels

Instructions:

a. In three of the mason jars, fill it with water.

b. Add red food coloring to one, blue food coloring to another, and yellow food coloring to the last container.

c. Here is a good time to discuss how to combine colors to make other colors. Example: yellow plus blue makes green and red and blue combined make purple.

d. After discussion, take all six of the mason jars and put them in a circle. Place an empty jar between each of the colored water mason jars.

e. Take the towel and twist it.

f. Put one end in the jar that has colored water and the other end in the empty jar.

g. Complete this until all jars have paper towels in them.

h. Allow 48 hours to pass until. The process will start immediately, and you will start to see the paper.

i. The end result will show the creation of new colors that will mimic a rainbow.

Let's Build a Robot!

FYI: This will not be your traditional robot.

Tools needed:

a. toilet paper or paper towel tubes

b. aluminum Foil

c. pipe cleaners

d. plastic straws

e. crayons

f. scissors

g. tape/glue

Instructions:

Use the tools above to create a robot. Be inventive and use this time to discuss how your student will build the robot, what it will look like, etc.

Let's Play a Game!
101 & Out

Tools Needed:

a. dice

b. paper

c. writing utensil

Instructions:

a. This game depends on the age of your student and can be altered according to their math competency level.

b. This can be done as one large group or multiple small groups (3 to 4 people per group). If done in small groups, each group will need their own die.

c. The goal is not to exceed 101 in your calculations (this can be 11, 21, etc. depending on the age of the group).

d. One student (or teacher) will roll the die. Whatever number the die lands on can be taken as that number or a multiple of ten.
    Example: the dice lands on a 6, the student can choose 6 or 60.

e. The student will write down the number and then wait for the next roll.

f. The next number is then added to the student's paper. They can either add or subtract number (multiplication and division is allowed as well).

g. If a person goes over, they are considered out until the next round.

Time to Write

Talk about something new you learned.

Talk about something that was interesting to you.

What topic do you want to know more about?

Talk about your favorite Woman in STEM.

Name some STEM activities you have tried lately.

What new STEM activities do you want to try?

Journal Pages

Journal Page

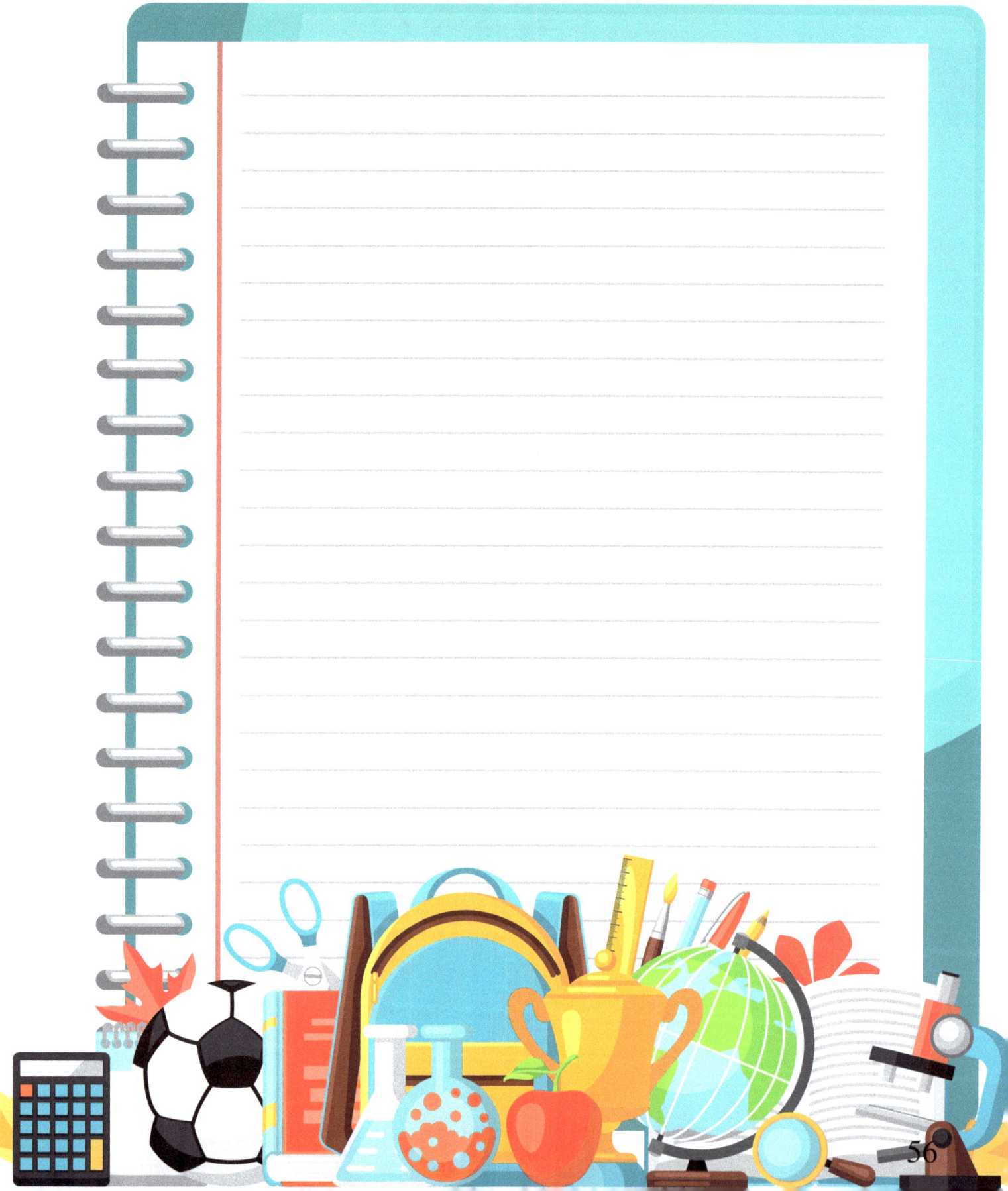

Journal Page

Journal Page

Journal Page

Journal Page

Journal Page

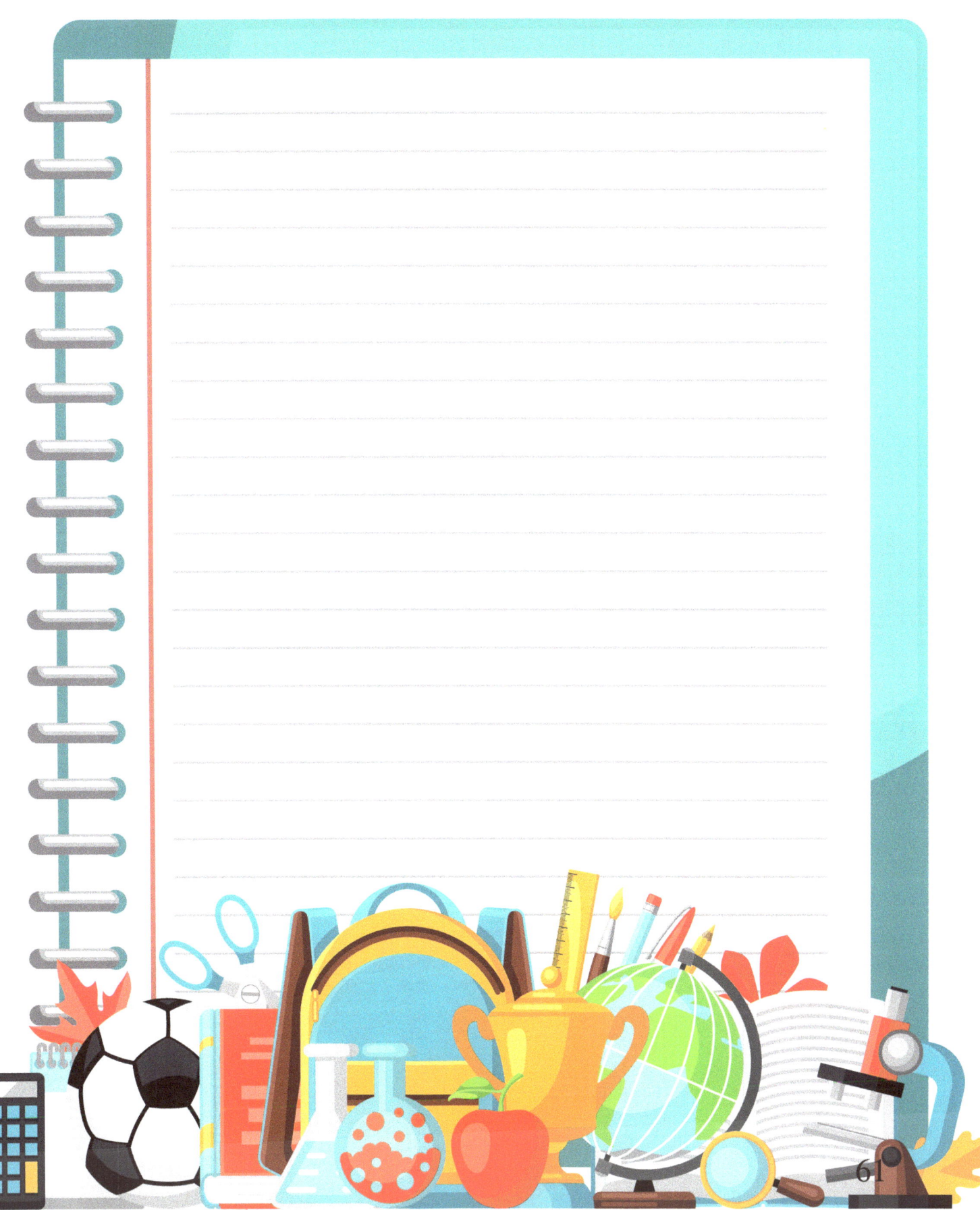

Journal Page

Journal Page

Journal Page

Solutions

Solutions

Match-Up Puzzle

Invention	j
Patent	h
Science	c
Experiment	m
Computer	n
GPS	g
Periodic Table	b
Three Dimensional	o
Data	l
Astronaut	e
Mathematician	d
Mechanic	i
Spaceship	k
Planets	f
Computer Science	a

Solutions

Number Block Puzzle

1

					7
					15
					15
0					9
			2		6
	2		2		16
12	12	13	12	12	14

2

					14
	2				4
					12
					7
4			3		14
4					16
13	7	9	15	9	10

Solutions

Number Block Puzzle Con't

3

3	4	3	5	0	**11**
1	1	3	4	4	15
1	1	0	3	3	13
3	**5**	0	4	**5**	8
2	5	0	4	5	17
					16
10	**16**	**6**	**20**	**17**	**13**

4

					12
0	5	5	4	**5**	19
2	0	1	1	5	9
5	1	3	2	0	11
5	1	3	2	2	13
2	0	3	4	**1**	10
14	**7**	**15**	**13**	**13**	**6**

Solutions

Hidden Message

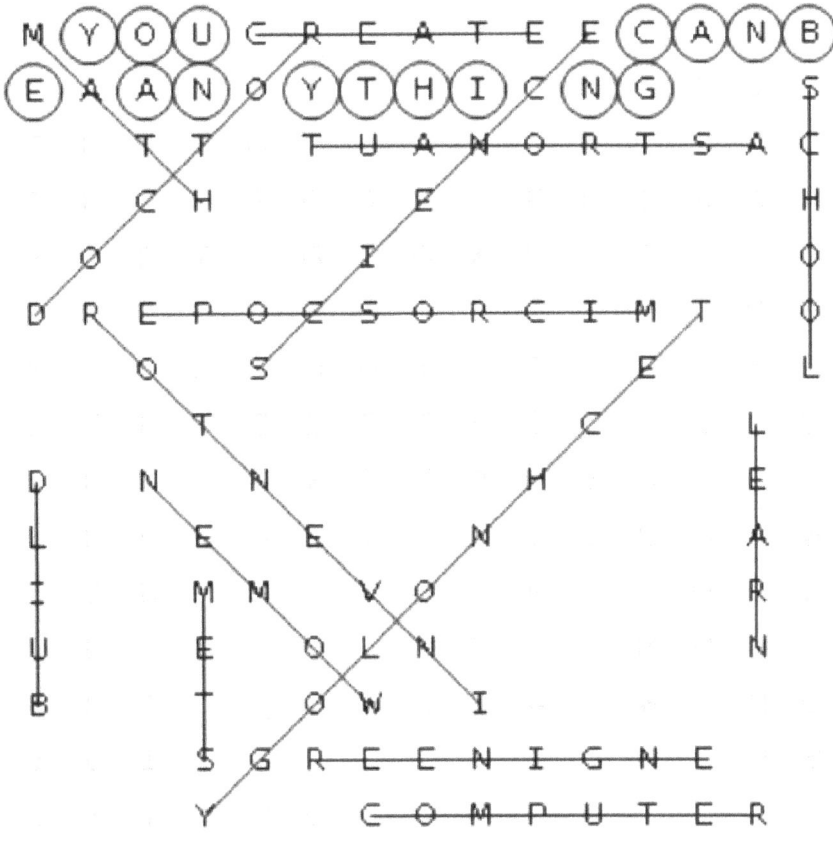

astronaut	build	computer
create	doctor	engineer
inventor	learn	math
microscope	school	science
STEM	technology	women

— — — — — — — — — — — — —

Solutions

Find the Word

1

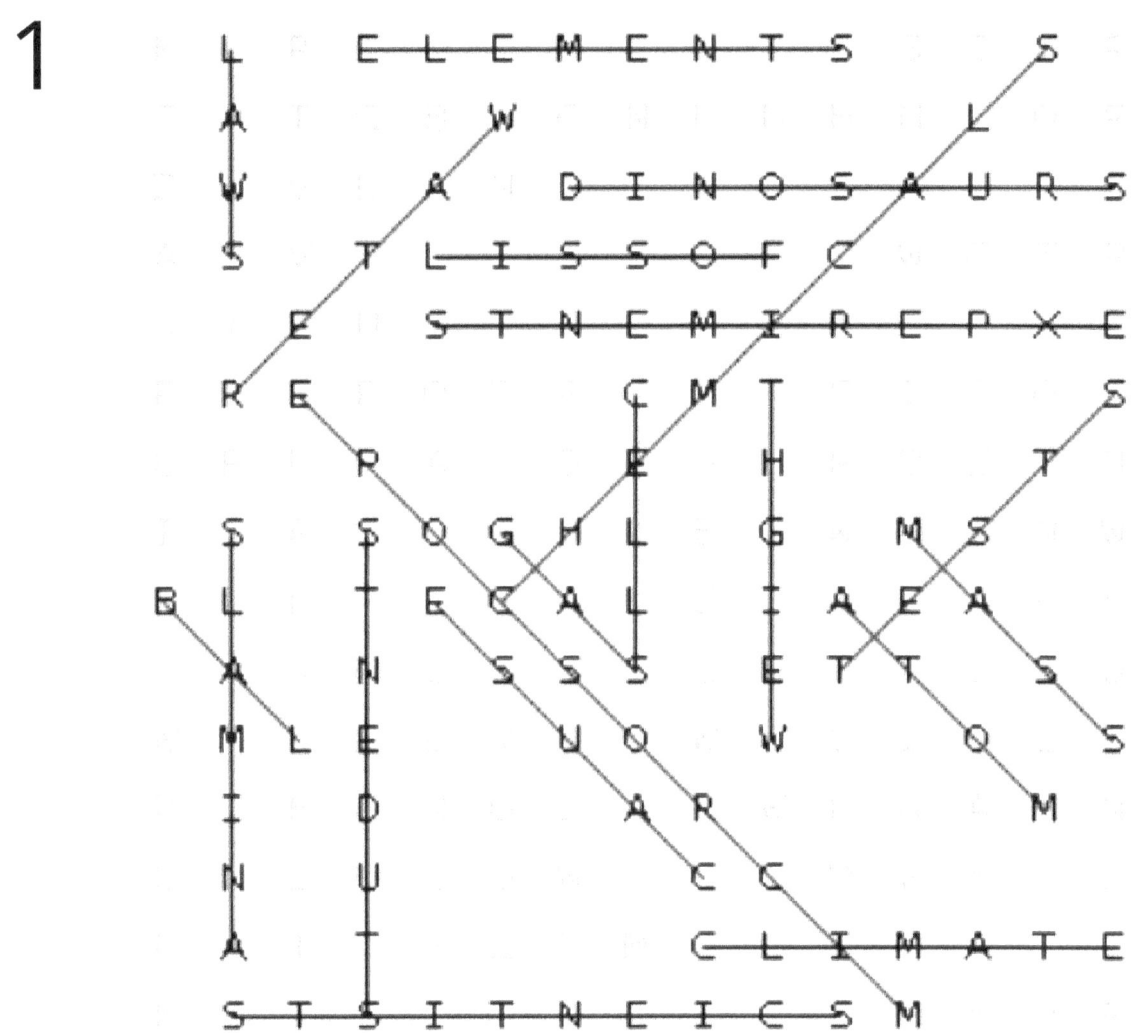

Solutions

2

Solutions

Find 5 Differences:

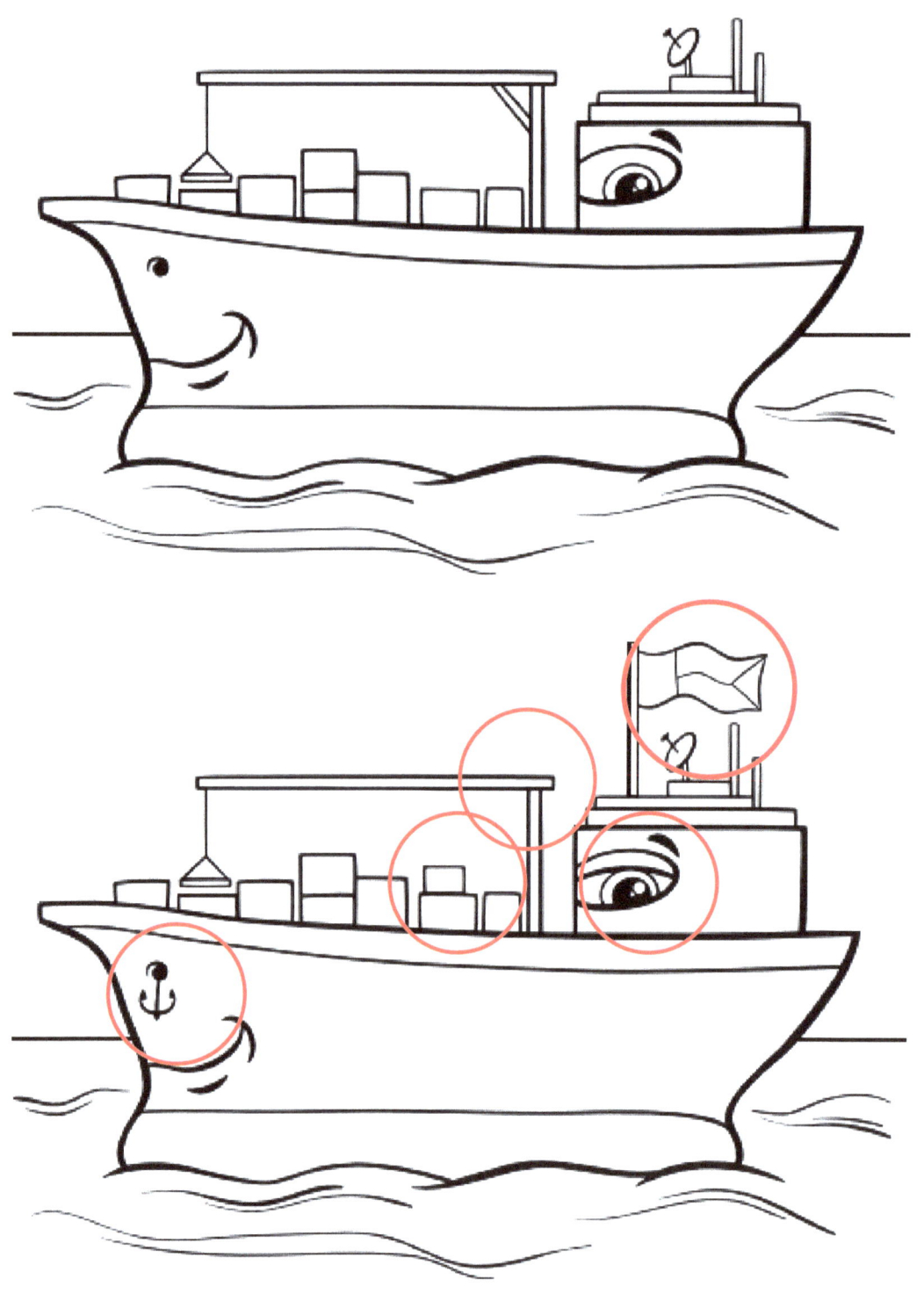

Solutions

Planet Crossword

Across

1. Saturn

3. Pluto

5. Venus

7. Mars

9. Neptune

Down

2. Jupiter

4. Earth

6. Sun

8. Uranus

10. Mercury

The End